Math Concept Reader

A Day at the Snack Stand

Printed in Mexico

ISBN 13: 978-0-15-360491-1
ISBN 10: 0-15-360491-3

1 2 3 4 5 6 7 8 9 10 039 16 15 14 13 12 11 10 09 08 07

Math Concept Reader

A Day at the Snack Stand

by Sarah Mastrianni

Photographs by Patrick Espinosa

Chapter 1: Oak Grove Park

The children in Oak Grove like to play soccer. Soccer teams practice during the week and play games on Saturday. Many fans watch games at the park. They love to clap and cheer. Sometimes it seems like the whole community comes out to see the teams play!

There is a snack stand at the park where players and fans can buy food and drinks. There are many healthful choices at the stand. The snacks give players energy they need to play hard. Cold drinks taste great to thirsty players.

Volunteers help at the snack stand. Today the Lee family is volunteering. The Lees will work together, and each of them will have an important job. This will help make the day go smoothly. The snack stand is always a busy place on game days. The Lees want to be ready.

The Lees are very excited about working. They take some time to talk about the jobs they will do. Mr. and Mrs. Lee will set up the snack stand. Sue and Jade will take and fill orders for customers. The Lees can hardly wait to get started.

Chapter 2: Getting the Snack Stand Ready for Customers

Mrs. Lee arrives at the snack stand early. She will help get food and drinks ready. People will arrive for the games soon, and Mrs. Lee wants to be sure they can see the snacks for sale. She will place juice boxes and granola bars at each customer window.

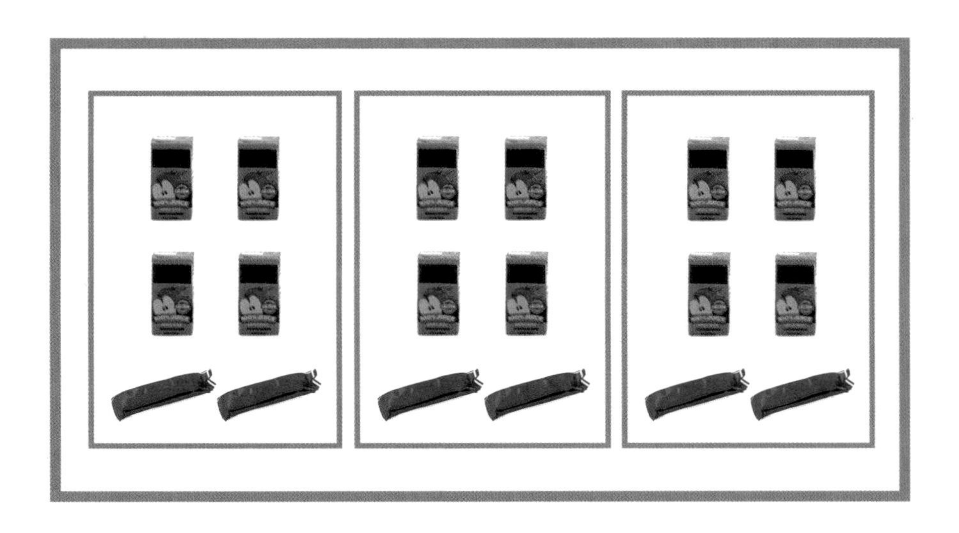

Mrs. Lee has 12 juice boxes. She divides 12 boxes into 3 equal groups. Each group has 4 boxes in it.

Mrs. Lee has 6 granola bars. She divides the 6 bars into groups of 2.

She places juice at each window. Then she places granola bars at each window.

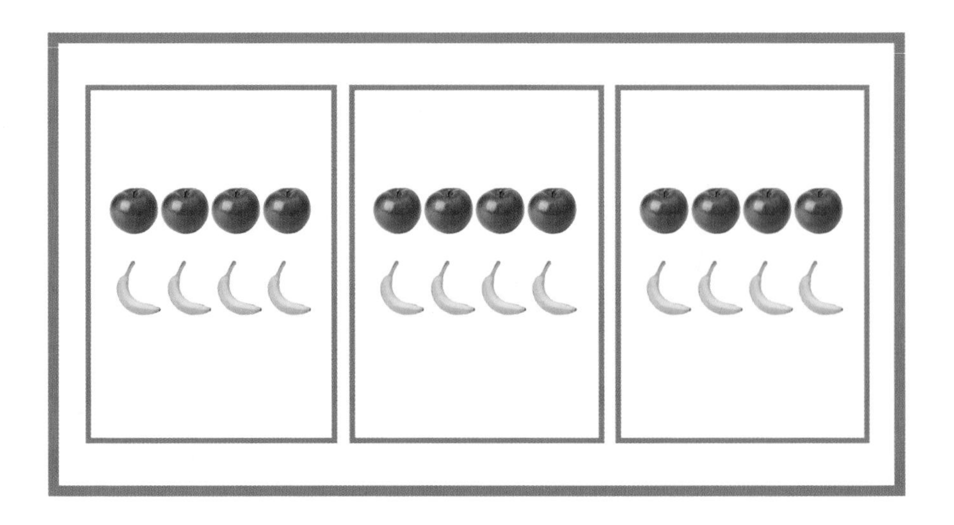

Mrs. Lee has some fruit to put out as well. She has apples and bananas. She divides the apples into 3 equal groups. Each group has 4 apples in it. Then she divides 12 bananas into 3 equal groups. Mrs. Lee puts 4 apples and 4 bananas at each window.

Mr. Lee arrives at the snack stand and opens the refrigerator. He sees many bottles of water. Mr. Lee will count the bottles on each shelf. He will keep track of the number of bottles they sell today. The weather is warm so he is sure they will sell many bottles of water.

Mr. Lee carefully counts the bottles of water in the refrigerator. There are 4 rows of bottles on each shelf, and each row has 7 bottles. Mr. Lee multiplies the number of bottles in each row by the number of rows.

He finds the total number of bottles on each shelf.

$4 \times 7 = 28$

There are 28 bottles on each shelf.

There are 5 shelves. He adds.

$$
\begin{array}{r}
\overset{4}{2}8 \\
28 \\
28 \\
28 \\
+\ 28 \\
\hline
140
\end{array}
$$

The Lees have 140 bottles of water to sell.

Chapter 3: Soccer Makes Everyone Hungry!

Jade and Sue help at the snack stand, too. They work together at a window. Sue takes orders from customers. Then she fills the orders. Jade takes the money from customers, and then she gives them change. Sue and Jade do a lot of math. They make an excellent team.

Mrs. Todd comes to the window, and Sue and Jade greet her with a smile. Sue asks Mrs. Todd if she can take her order. Mrs. Todd orders an apple and a bagel.

apple	$0.50	
bagel	$0.75	

The apple costs $0.50. The bagel costs $0.75.

Sue adds to find the cost of Mrs. Todd's order.

Mrs. Todd's Order		
apple		$0.50
bagel		+ $0.75 = $1.25

The total order costs $1.25. Mrs.Todd gives Jade $2.00. Jade subtracts to find Mrs. Todd's change.

$$\begin{array}{r} \$2.00 \\ -\ \$1.25 \\ \hline \$0.75 \end{array}$$

Mrs. Todd will get $0.75 in change. Jade hands Mrs. Todd the coins. Sue gives Mrs. Todd her food. Mrs. Todd thanks them.

Beth, Jade's good friend, is next in line with her Dad. Beth has a big smile on her face as she says hello. Her game just ended, and she scored two goals. Her soccer team won its game today. Beth is very happy.

Beth is hungry for a snack. She thinks about what she would like to order. She chooses a sandwich and water.

sandwich	$2.35	
water	$1.75	

The sandwich costs $2.35. The water costs $1.75.

Sue adds to find the total of Beth's order.

Beth's Order		
sandwich		$\begin{array}{r} ^{1\ 1} \\ \$2.35 \\ +\ \$1.75 \\ \hline \$4.10 \end{array}$
water		

Beth's order costs $4.10. Beth gives Jade a five-dollar bill. Jade subtracts to find Beth's change.

$$\begin{array}{r} ^{4\ \ 10} \\ \$\not{5}.\not{0}0 \\ -\ \$4.10 \\ \hline \$0.90 \end{array}$$

Jade gives Beth $0.90 in change. Sue hands Beth her order. Beth says thank you. She walks to a picnic table to eat her snack.

Chapter 4: Everyone is a Winner

It is late in the day. The soccer games are ending. The sun is slowly going down, and people are leaving the park. The Lee family cleans the snack stand. They make sure everything is put away neatly. Mrs. Lee counts the money. There is one more thing to do.

Mr. Lee opens the refrigerator to count the drinks. He needs to order more water. People will need plenty of water for next Saturday's games. Mr. Lee thinks. How many bottles of water did they have when they opened the snack stand? How many bottles are left?

The Lees began the day with 140 bottles of water. There are only 22 bottles of water left. Mr. Lee subtracts.

Mr. Lee will order 118 bottles of water. This way, there will be enough water for next week's thirsty players and fans.

It has been a busy day. The Lees enjoyed working at the snack stand, and they hope to do it again soon. They made a great team.

Now it is time to go home. Mr. Lee locks the snack stand. They all agree that game day was great fun!

Glossary

add to join 2 groups

divide to place into equal groups

multiply to join equal groups

subtract to take away objects from a group or to compare groups

volunteer a person who does a job without getting paid